AF487835

THIS BOOKS BELONGS TO

A a is for Animals

Bb is for Bat

Cc
is for
Cow

Dd *is for*
Dolphin

Ee  *is for* **Egg**

Ee e e e

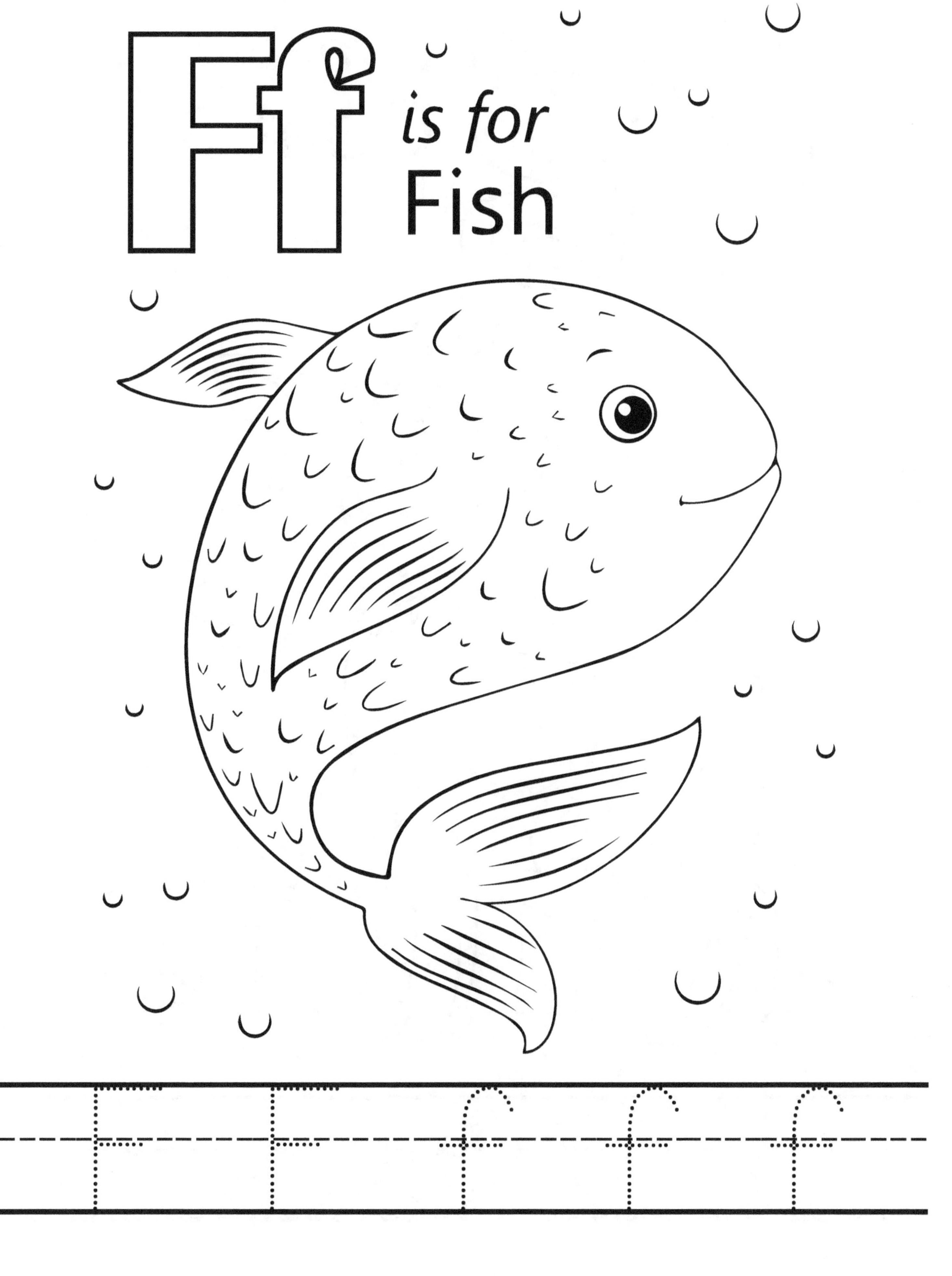

Ff
is for
Fish

G g is for
Goat

Hh
is for
Horse

Ii *is for*
Ice cream

Jj
is for
Jaguar

Kk

Ll
is for
Llama

M m *is for* Mouse

M M M m m m

Nn
is for
Nest
N N n n n

O is for Orange

P p

is for

Parrot

Q q is for Quail

R r is for Rabbit

S is for
Spider

Tt

Uu

V is for
Vase

W
is for
Worm

X is for Xylophone

Yy is for
Yak

Z

is for
Zebra 

Let's Start warming Up:

Trace the Number 0

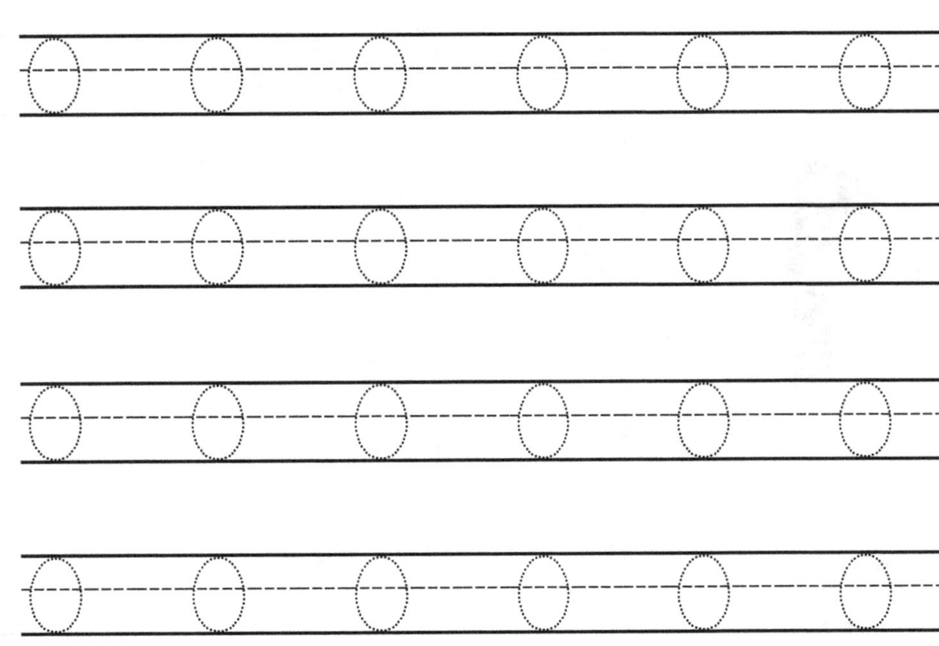

Trace the word Zero

Trace the Number 1

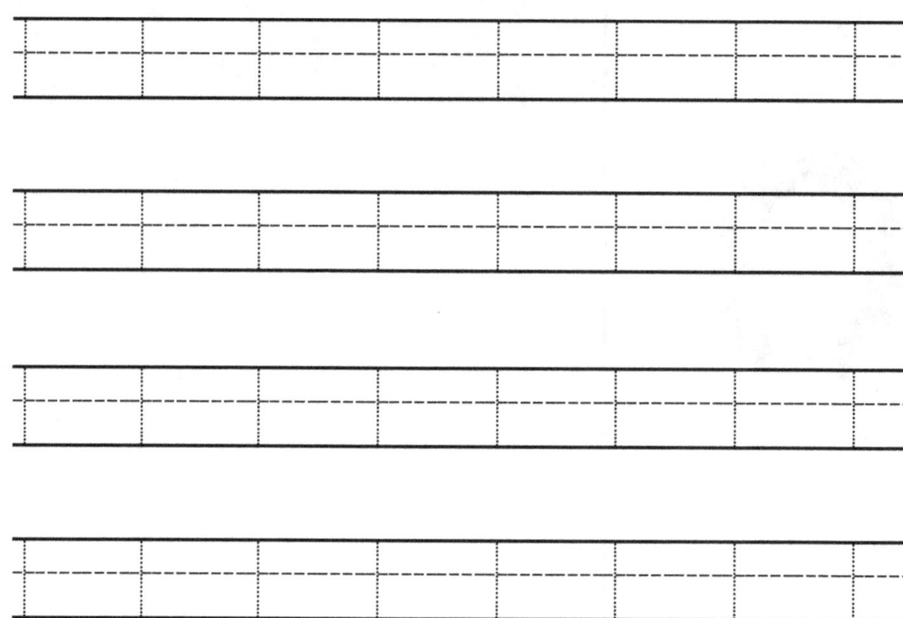

Trace the word One

1

Write Number 1

1
1
1
1

Color One Orange

One

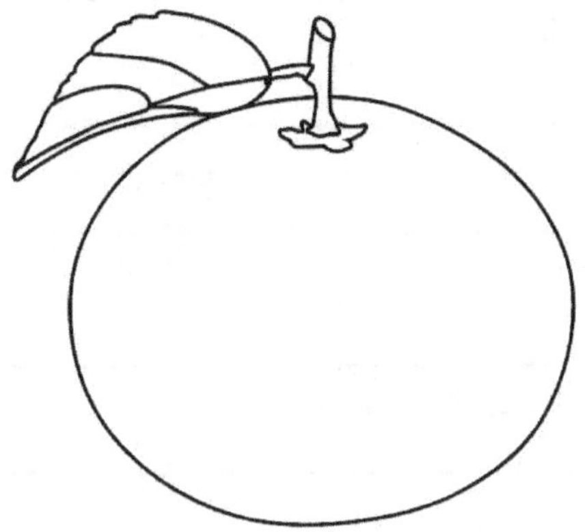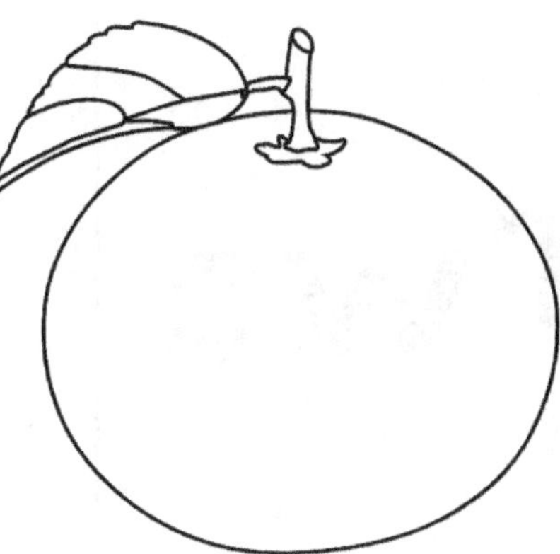

2

Trace the Number 2

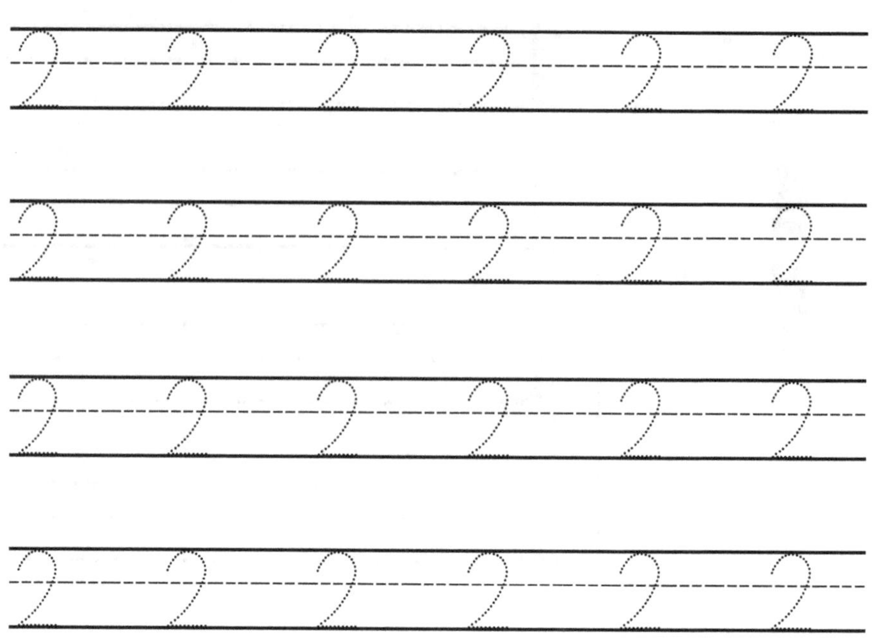

Two

Trace the word Two

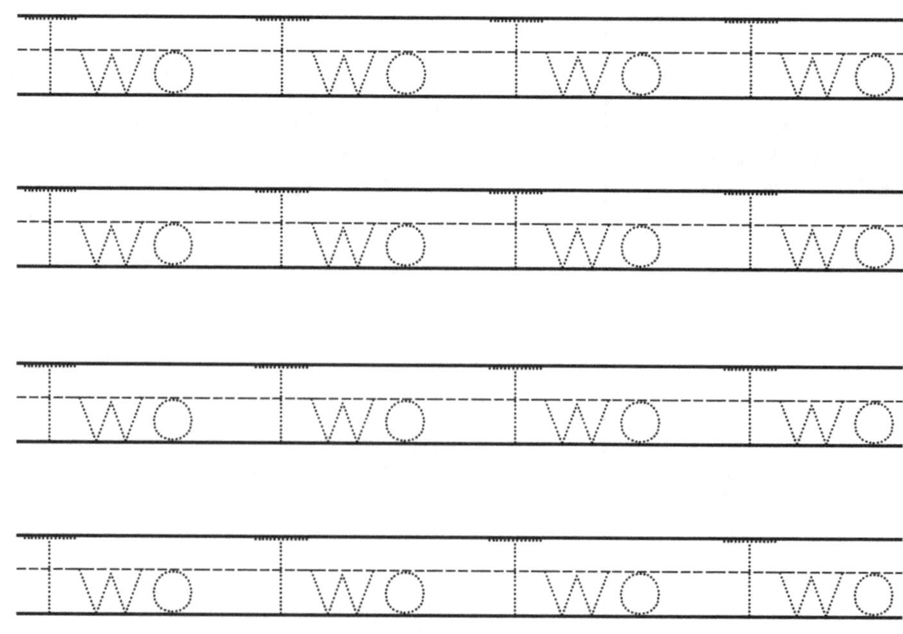

2

Write Number 2

2 ----------
2 ----------
2 ----------
2 ----------

Two

Color Two Apples

3

Trace the Number 3

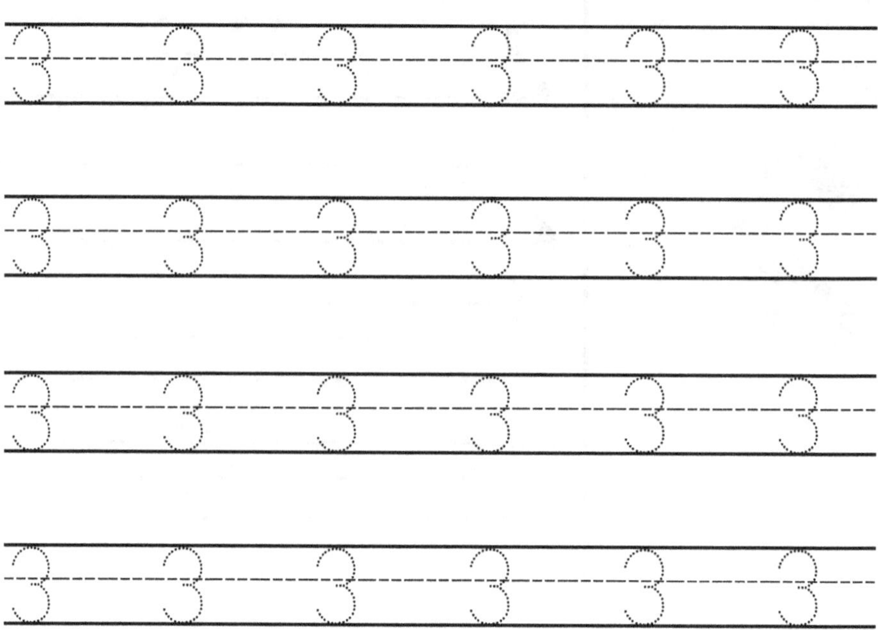

Three

Trace the word Three

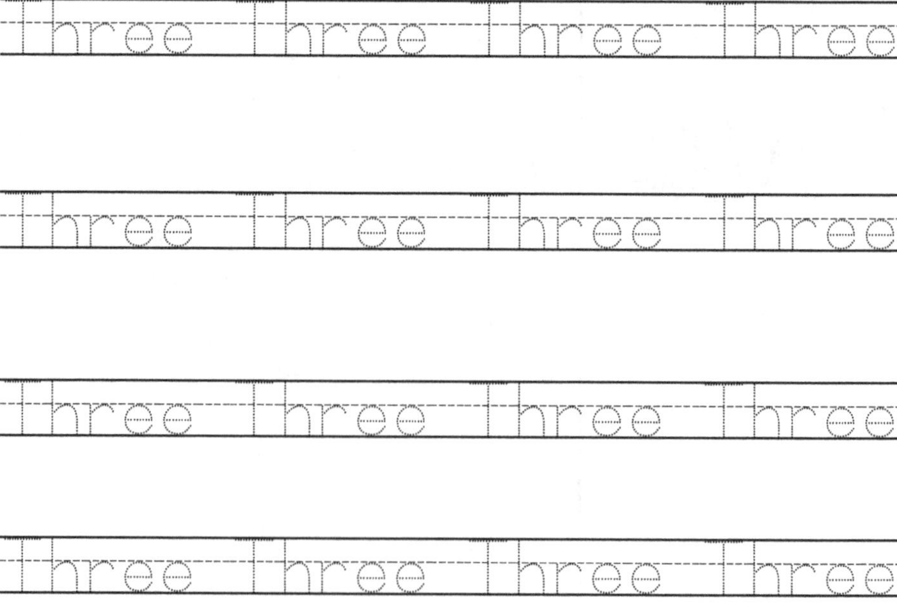

3

Write Number 3

3
3
3
3

Three

Color Three Pears

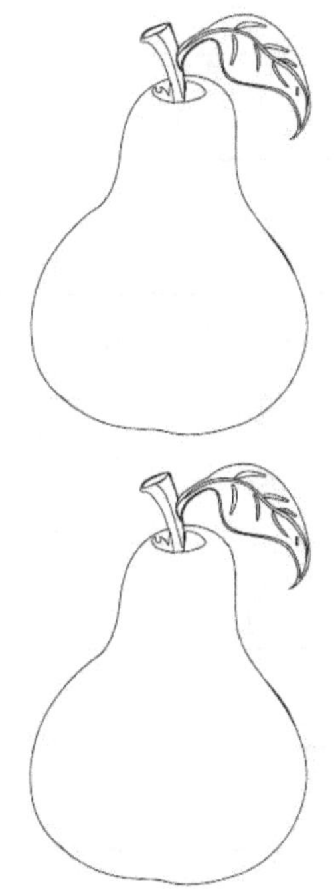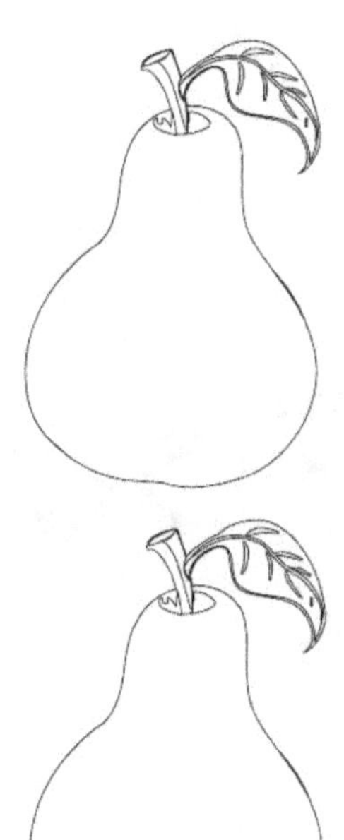

Trace the Number 4

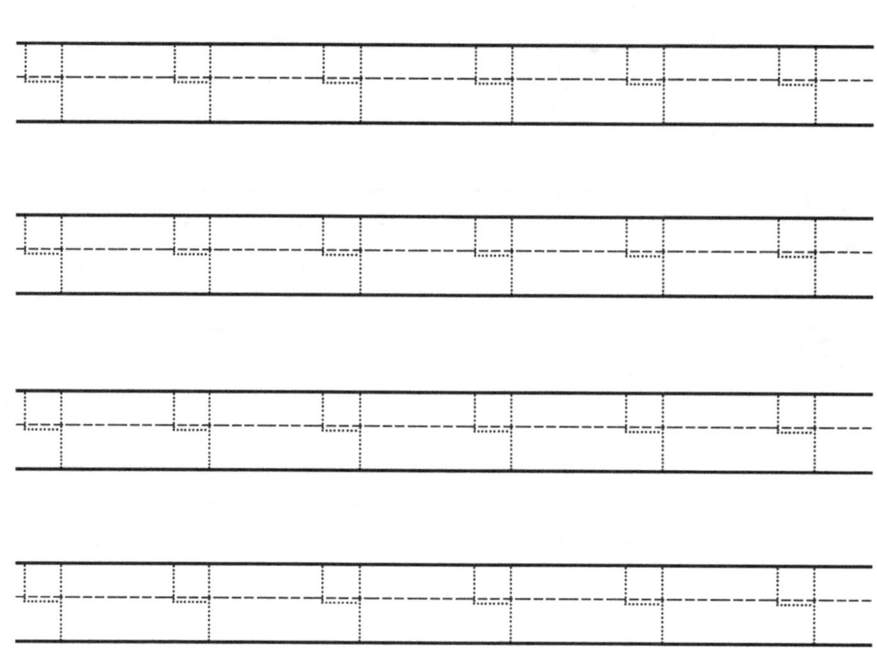

Trace the word Four

4

4
4
4
4

Four

Color Four water Melon slices

Trace the Number 5

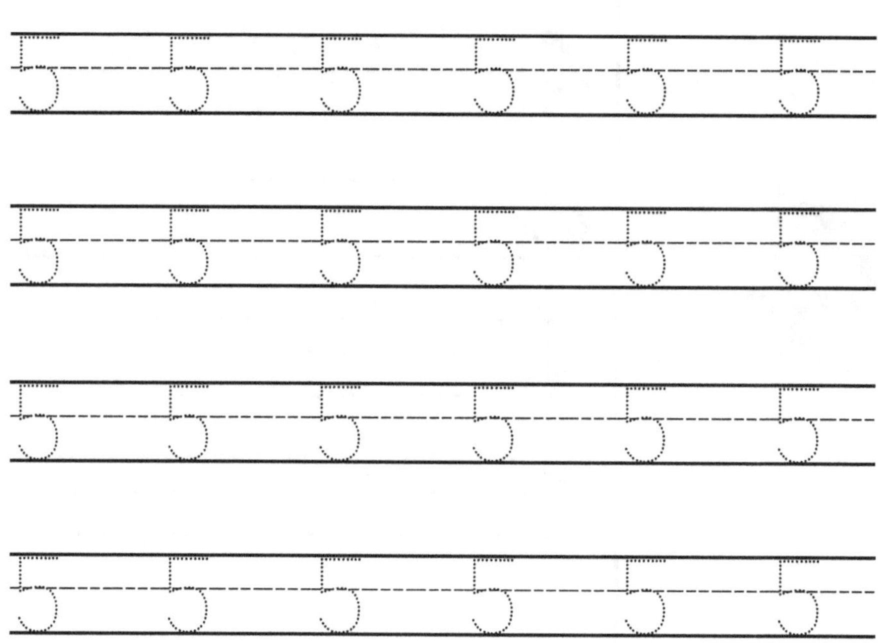

Trace the word Five

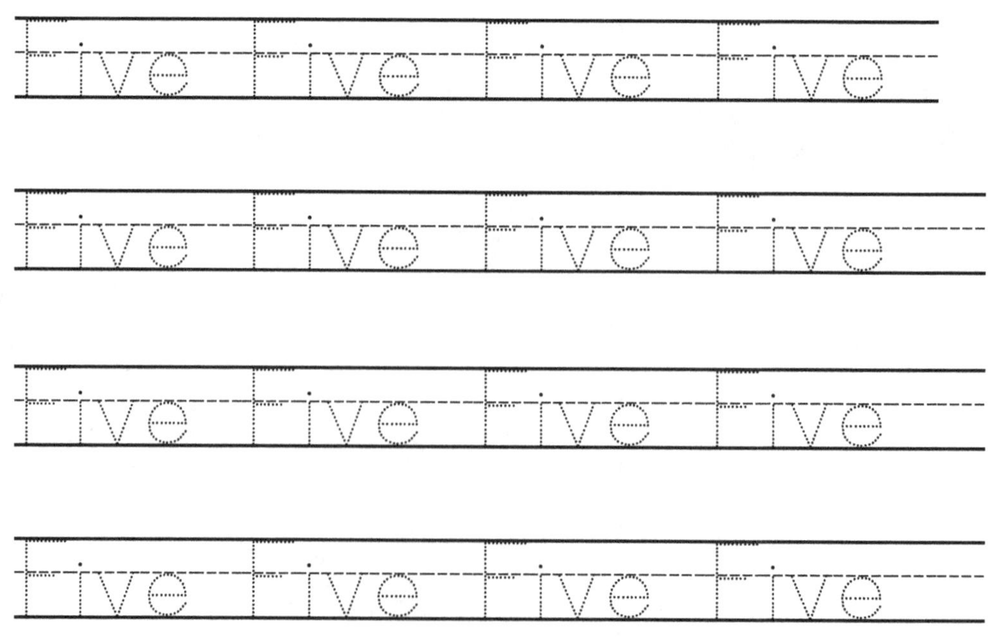

5

Write Number 5

Five

Color Five Tomatoes

Trace the Number 6

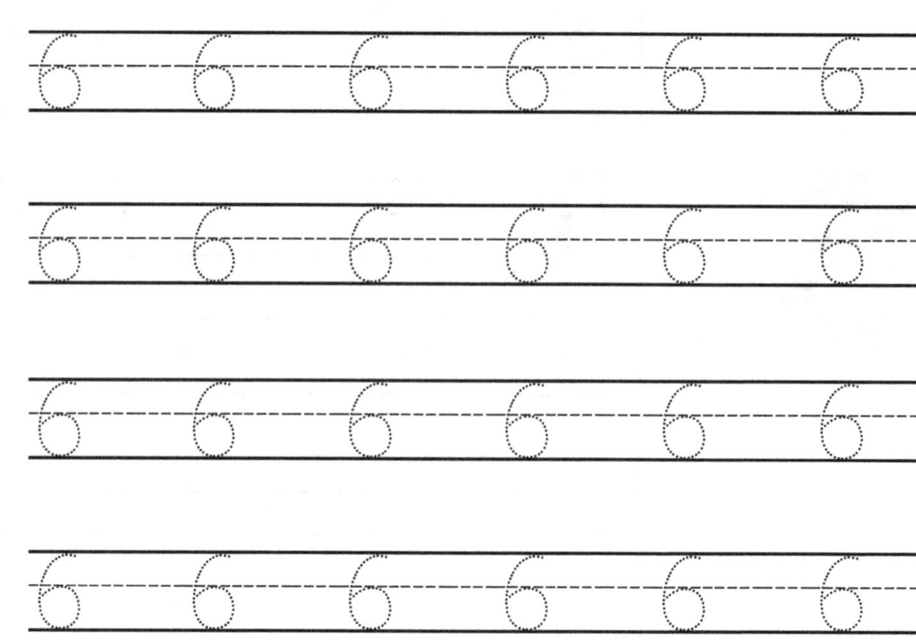

Trace the word Six

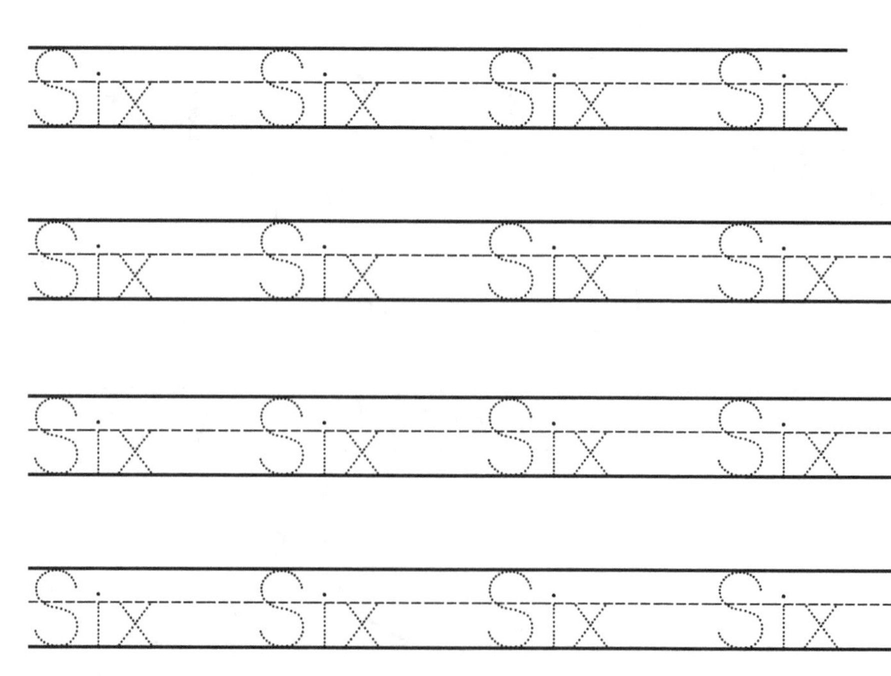

6

Write Number 6

6

6

6

6

Six

Color Six Carrots

7

Trace the Number 7

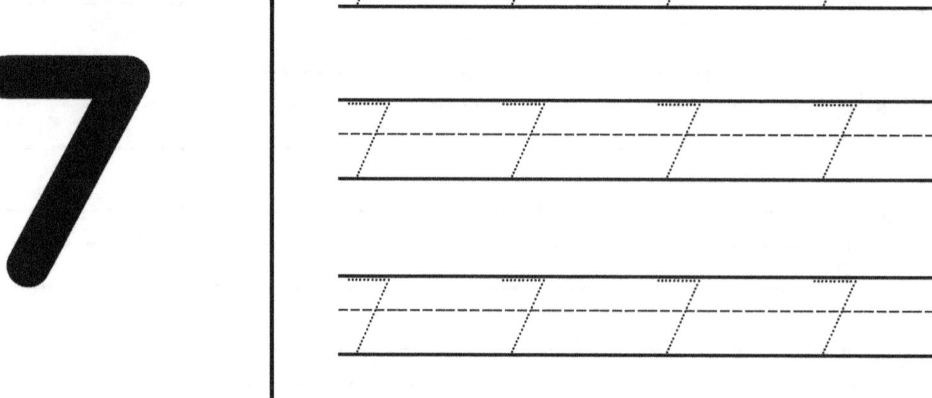

Seven

Trace the word Seven

7

Write Number 7

7 7 7 7

Seven

Color Seven Pumpkins

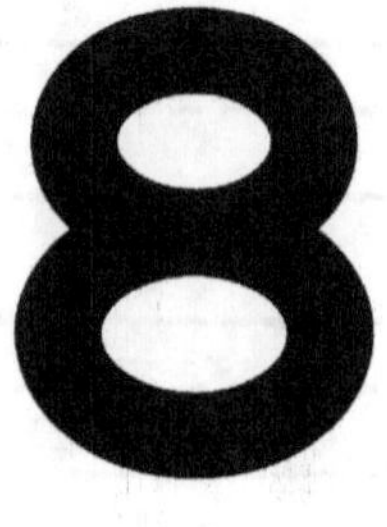

Trace the Number 8

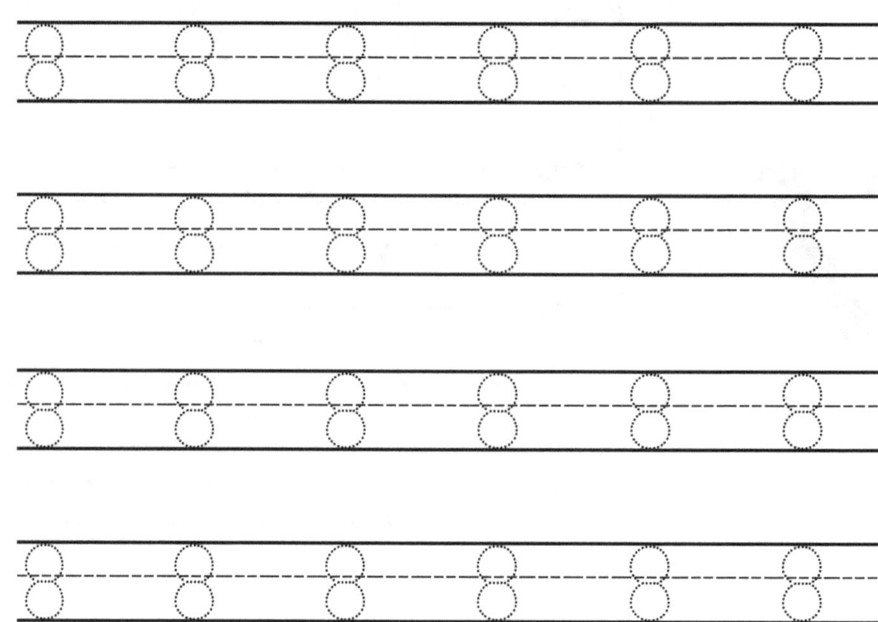

Trace the word Eight

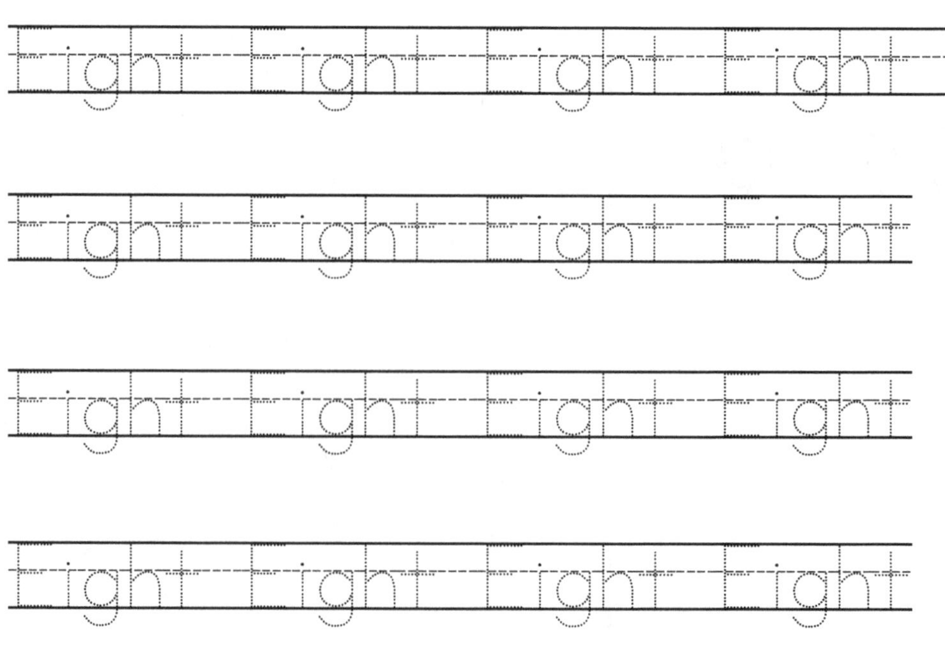

8

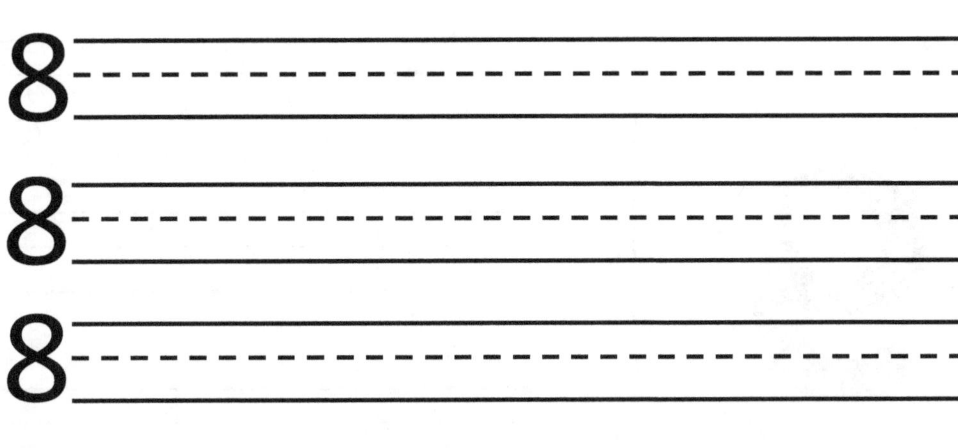

Eight

Trace the Number 9

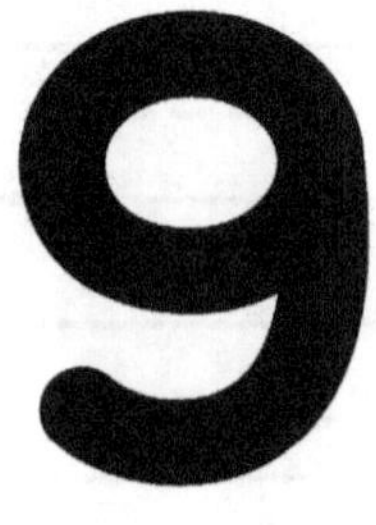

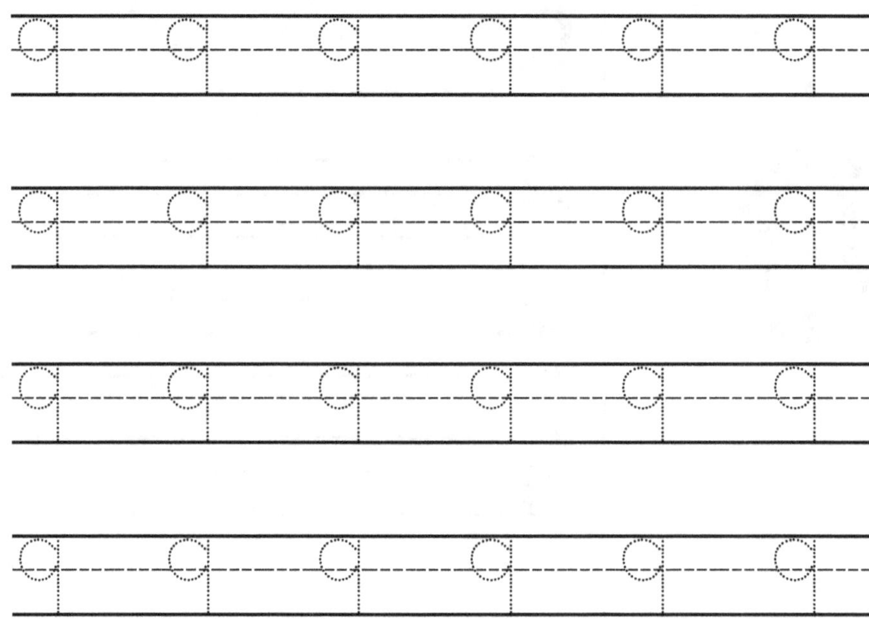

Trace the word Nine

9

Write Number 9

9 9
9 9
9 9
9 9

Nine

Color nine Spikes of wheat

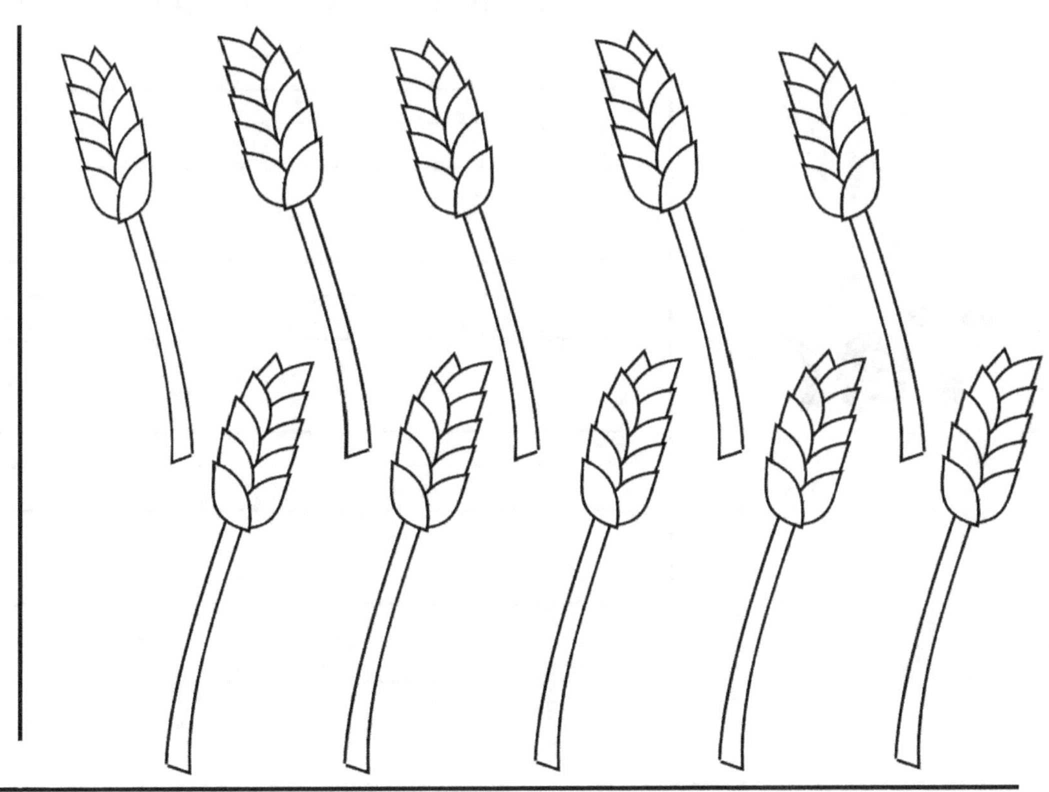

10

Trace the Number 10

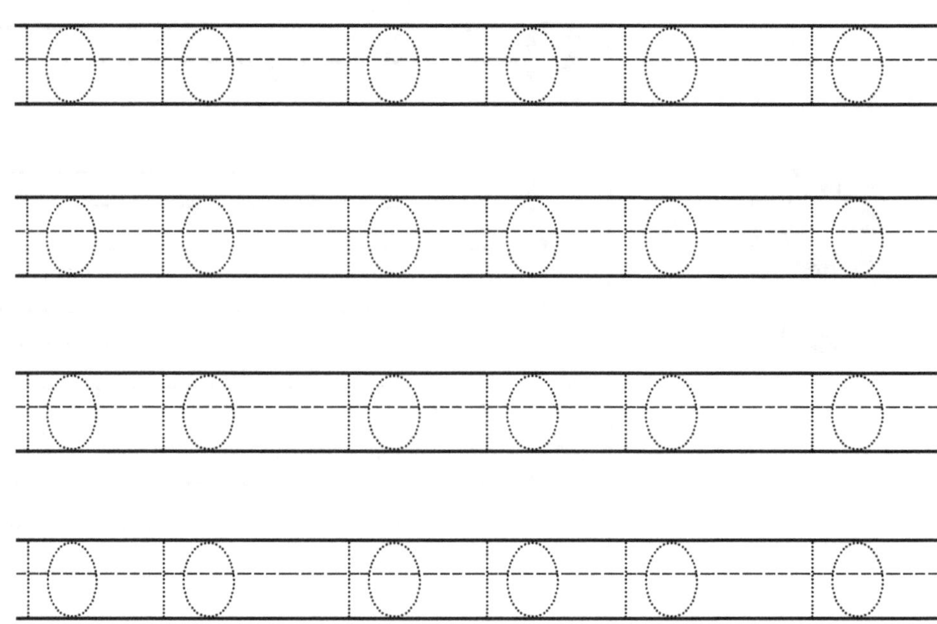

Ten

Trace the word Ten

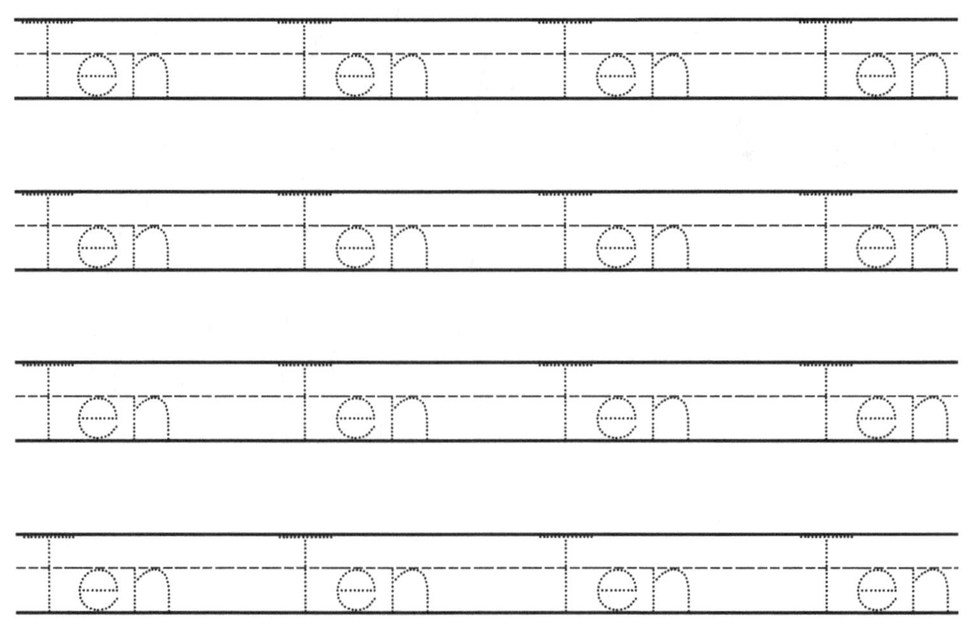

10

Write Number 10

10 10 - - - - - - - - - - - - - - -

10 10 - - - - - - - - - - - - - - -

10 10 - - - - - - - - - - - - - - -

10 10 - - - - - - - - - - - - - - -

Color Ten Cherries

Ten

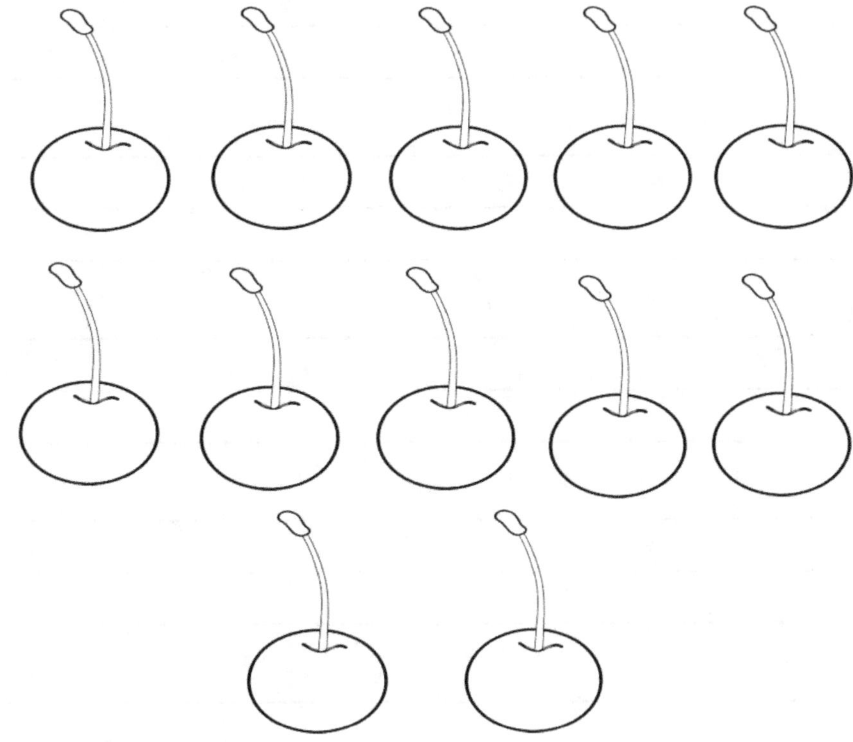